The C.Y.O 'Connor ERADE Village Foundation

Incorporating the Immunogenetics Research Foundation and the Institute of Molecular Genetics and Immunology Incorporated

E*ducation*, **R***esearch* **A***nd* **D***evelopment*, **E***mployment*

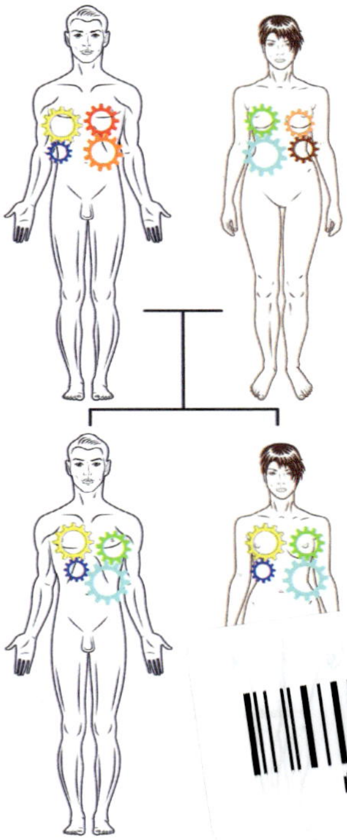

32nd Annual Report 2015

C Y O'Connor ERADE Village Foundation Inc.

Employment Research and Development
Employment Technology Incubator

Street Address	24 Genomics Rise Piara Waters WA 6112
Postal Address	PO Box 5100 Canning Vale South WA 6112
Telephone:	(+61 8) 9397 1556

Key Contact	Professor Roger Dawkins
Position	Chairman
Email	rldawkins@cyo.edu.au
Second Contact	Dr Sally Lloyd
Position	Director
Email	slloyd@cyo.edu.au

Company Overview : CYO 'Connor ERADE Village Foundation seeks to harness the intellect of young Australian post graduates whilst maintain extensive collaboration with leading institutes overseas.

Recently expanding into multidisciplinary science in fields such as bioenergy, information technology, natural products, recycling and water, the core technology is interrogation of genetic information.

The aims include understanding of human evolution in the context of disease including retroviruses, identification of new genetic markers via genomic prospecting and development of diagnostic tests for inherited traits in numerous species of medical and commercial interest such as cattle, dogs, fish and invertebrates.

Product/Service Information: The CYO'Connor Education Research and Development Employment (ERADE Village) Foundation is a charitable Foundation that conducts independent and sponsored research and administers a federally funded technology incubator.

The Foundation has available low-cost two story R&D studios including accommodation if required, central laboratory facilities, conference rooms, visitor accommodation and sophisticated intranet and broadband connection.

Fellowships are awarded to outstanding researchers to assist them in commercialisation of their R&D.

Industry and government funding have established the Village as a multimillion dollar R&D and commercial centre following a model that is being replicated at the Albany Research Institute

Published by Nearurban LLC, 1603 LBJ Freeway,
Suite 280, Dallas, TX 75234
Tel (972) 974 7130, Fax (214) 393-4001
Email : admin@nearurban.net

Manufactured and printed by CreateSpace.

ISBN: 978-0-9864115-2-6

Title page image: From R.L.Dawkins' "*Adapting Genetics – Quantal Evolution After Natural Selection-Surviving The Changes To Come* " 2015 Nearurban Publishing ISBN 978-0-9864115-1-9 : **Reactive meshing dependent on hormonal and other environmental influences.** Coloured cogs represent ancestral haplotypes. The size and shape of cogs represent polymorphism or variant forms of the blocks. The two offspring are genotypically identical but the interactions are different as a consequence of the hormonal environment.

NU Publishing

FRIENDS OF CYO'CONNOR VILLAGE

The Foundation is a charitable organisation supporting education , research and development and employment. Without the help of friends of the Foundation it would not be possible to achieve all these objectives, including exciting new scientific discoveries. The Board thanks its many friends for their generous donations, bequests, collaborations, advice and assistance during 2014.

(Cut here)

- -

I/we (Name/s) ....(Tel)................

Of (Address)..wish to:

Join the "Friends of CYO" 2015 ($55.00 pa) $.................

Receive information about the Corporate Golf Day and other fundraising events ......

Be invited to Scientific Retreats ......

Receive information about arranging a bequest ......

Support the Foundation in other ways ......

Please accept my / our total payment of $.................$

Cheques/Money Orders payable to the : **CYO'Connor ERADE Village Foundation** or,

Please debit my **Credit Card** with the above total amount only. I understand that American Express, Diners, Mastercard or VISA are accepted. My card details are as follows:

Card Type: ....
Card No.: .... Expiry Date ..../,,,,,,,

Cardholder's Name : .... Signature ....

POST TO : THE C. Y. O"CONNOR ERADE VILAGE FOUNDATION INC.
PO BOX 5100 CANNING VALE SOUTH WA 6155

AL & M DAWKINS FOUNDATION

Sponsors up to

$200,000 p.a. for Research & Development in WA

Residential scholarships and grants for:

FUTURE OF WA
BIOMEDICAL RESEARCH

in association with a university within WA if enrolled. Matching funding must be available or anticipated. Expressions of interest including track record should be addressed to:

CY O'Connor Village,
PO Box 5100, Canning Vale 6155

Closing Date

November 30th or May 31st each year

The AL & M Dawkins Foundation has been funded to support problem solving in Western Australia by:

(1) Defining important issues relating to urban development and planning and potential solutions thereto;

(2) Defining important issues relating to biomedical research and solutions thereto;

(3) Supporting prospective innovators in the above fields and by funding awards, scholarships, prizes and fellowships in association with CY O'Connor ERADE Village, or any branches thereof, in association with any university to which the CY O'Connor ERADE Village may be associated;

(4) Provide funding for patent costs to protect any intellectual property so developed.

Applications received by November 30 or May 31 each year will be evaluated by one of two expert panels which make recommendations to the Advisory Board,

The Board may seek further advice and will also consider:

(1) the relevance to multidisciplinary projects underway within the CYO

(2) the value of the individual to life within the village

(3) the capacity to leverage other funds

(4) university linkages

(5) potential IP

and will offer funding and accommodation for live-in researchers.

The town planning panel will be seeking to support innovative projects relating to water shortages and contamination, affordable housing, sustainable villages and isolated communities.

The biomedical panel will identify boutique projects which can be internationally competitive whilst relevant to local issues such as DNA matching, evolutionary genetics, livestock diseases and plant genomics.

Within 3 months, the administrators of the fund will advise applicants in the following categories

(1) successful

(2) for reconsideration when matching funding is confirmed

(3) invitation to reapply in the next cycle

(4) unsuccessful

Grants, generally of $50, 000 must include a stipend and rental but may also include funds for technical support, equipment, reagents, patents and publications. Travel grants will not be considered. The Foundation may decide to make no awards.

Funds will only be released after assignment of IP in return for a royalty negotiated on a case by case basis.

Table of Contents

Patron

Mr. S.L. Perron AM

Vice-Patrons

The Hon. Kim Beazley MA, MPhil
Senator Alan Eggleston BA, MRCS, LRCP, FRACGP
Mr. Max McDonald
Mr. Patrick Dawkins
Mr. Syd Martin

Board of Management

as of June 2014

Appointed Members

Emeritus Professor Roger L. Dawkins (MD,DSc,FRACP,FRCPA,FRCP)	- Research Director/Chairman
Mr. Anthony Lloyd (BSc (Ag)(Hons)MBA,LLB)	- Vice Chairman
Vacant- Dr. Joseph Williamson (BSc (Hons) PhD) – Redundant October 2014	- Deputy Research Director
M. Renny Lee-Steere, (EMBA)	-Deputy Vice Chairman
Mr Edward Turner (FPIA, MTP, ATARP)	- Honorary Secretary
Ms Sue Lester (BSc)	- Deputy Honorary Secretary
Mr. Wayne Cox (MTax, DipFS(FP), CPA)	- Honorary Treasurer
Mr Jason Wallis (BBus, R.E.Lic)	-Deputy Honorary Treasurer

Nominations for Elected Members

Mr Ben Dawkins (LLB)
Mr Roger Stubbs (AM BA, Grad Dip URP)
Dr Sally Lloyd (BSc, PhD)
Mrs Sharon baker – Appointed May 2014

Co-opted Members (by Invitation)

Dr Sam Dawkins, BSc, PhD
Mr Brian Walker

Life Members

Dr BJ Stewart, BVSc (Hons), Dip Ag
Mr John Heydon, BCom, CA, FTIA
Mr David GA Gibb, BA (Hons), AIBMS, AAIMS
Mr Lindsay Wolridge

Chairman's Report

Professor Roger L. Dawkins

Nearurban Reports

After 33 years of adaptation, the CYOEVF now introduces a major change to the Annual Report. This year we will utilize the services of *Nearurban Publishing* to provide a new print format and an electronic version which can be read on your computer or Kindle reader.

This development has been driven by Ted Steele who kindly mastered the intricacies in the course of publishing two books for us. He has been assisted by Peter Hinchliffe who was responsible for the desktop publications of earlier years.

There will be plenty of experiments: We look forward to your active involvement in this venture.

The Plan

The main focus for the Board has been preparing for the move of the CYOEVF from urban sprawl to a Greenfield site in North Dandalup and the proposed "Food Bowl".

The motivation is quite simple. The Foundation needs to strengthen its contacts with producers and especially those committed to innovation in the application of our skills to the success of the area. We need to recruit bright youth whilst utilizing older folks to create an interesting and hopefully exciting future for CYOEVF.

We see an urgent need for a Research and Development Centre which, in keeping with the ERADE principle, provides education and encourages employment. The laboratories will need to be multidisciplinary and adaptable with the ultimate aim of distributing the technology and the actual testing to the producers.

Particular emphasis should be given to :

1. Preparing for change or "future proofing" the beef industry by preparing for,
• Increasing insistence on health (low melting temperature fat on grass),
• Assurance of provenance from paddock to plate (simple, cheap, DNA testing available to the producer),
• Improving the quality of northern live and packaged beef by judicious genetic blending with seed stock which is now exported to international competitors (producer specific solutions),
• Real or contrived trade barriers based on fears of disease.

2. Training of those already committed to the industry.
• Competence in the above,
• Technical capacity in livestock genetics and reproductive biology,
• Decision making in the presence of complexity and misleading advice from industry bodies.

To these ends, the foundation has appointed Knight Frank to manage the existing incubators with a view to selling when optimally renovated and leased over the next 2 years. The proceeds become the initial capital fund for the new buildings which, with fit-out, are estimated to cost $10m.

CYOEVF fully understands that the staffing and operation of the buildings will require a capital fund of the same order.

It follows that the project depends upon land and servicing being provided from other sources. Therefore no commitments will finalized until the site is secure and functional.

Over the next several months we will be seeking members of the initial council to select the appropriate site from those available. The ALMDF via CYOEVF has allocated a contribution to the salary of an executive officer who will assist the council.

Research and Development

As elaborated in subsequent articles, we can be proud of progress on several fronts.

Genetic testing has become the major focus and has now been absorbed in a national collaboration with leading Healthy beef (Wagyu) producers including Scott de Bruin in South Australia, the Hammonds in Tasmania, Peter Gilmour in WA and, of course, Dom Bayard in Victoria. This is a long term project currently funded entirely by CYOEVF. The politics at all levels are entrenched as reflected by the automatic rejection of our offers to be involved with Meat and Livestock Australia.

The reality is clear. Australian agriculture needs more and better R&D. For example CYOEVF has demonstrated the unequivocal utility of its patented technology in correcting pedigrees based on the imported technology.

The healthy beef project has flourished with the development of the patented low Tm approach. As we show through the blind tasting programme, healthy beef is also tasty beef. Low Tm steak is delicious whilst also lowering cholesterol and reducing the requirement for statin therapy which in Australia costs more than a billion dollars per year. This is important since earlier clinical trials of low fat diets failed through lack of compliance. The participants would not eat low fat meat. With new trials in mind, Sally Lloyd presented our approach to the Wagyu field day hosted by Peter Gilmour.

Publishing

On behalf of *Nearurban Publishing*, Ted Steele has finalized the publication of two books:

Ancestral Haplotypes : Our Genomes Have Been Shaped In The Deep Past. By Edward J Steele , 2015 Nearurban, Dallas, TX ISBN 978-0-9864115-0-2. 203 pages

The 33yr history of the foundation is told through its retreats, publications and discoveries. He quotes Winston Churchill: "The truth is incontrovertible. Malice may attack it, ignorance may deride it, but, in the end, there it is."

Adapting Genetics: Quantal Evolution After Natural Selection – Surviving The Changes To Come. R. L. Dawkins 2015 Nearurban, Dalla, TX. 978-0-9864115-1-9. 173 pages

As the cover says: "Genetics and evolution, as taught for a century, imply that survival is a lottery dependent upon random errors in copying the genome. ***Quantal evolution*** refers to the partitioning of the genome to protect critical regulatory DNA sequences, including during the successive ***retroviral*** epidemics which defined human ***ancestors***. Trial and error selected pre-tested combinations which are faithfully inherited thereby allowing the new species to adapt by changing gears in the ***absence*** of natural selection or mutations. Heterozygous ***polymorphisms*** interact polymorphically, thereby creating additional options. There are clear implications for ***forensics***, ***livestock selection*** and ***transplantation*** inter alia. Anamnestic evolution ensures individuality, diversity, protection from repeat epidemics and survival in the future."

Ted has done us all a great favour in bringing cost-effective international publishing to CYO and its collaborators. Send him proposals ranging from scientific treatises through memoires and annual reports to best selling racy novels.

Retirement

This year was most memorable for the retirement of Joe Williamson. I worked with Joe for 40 years and, more than anyone, I appreciated his meticulous but inspiring approach to the laboratory. I can think of close to 100 colleagues and students who depended upon him to fine-tune their skills to the level they needed to obtain their positions, grants and theses - and these are only the locals. Add the setting up of techniques in the labs of our friends such as Jim Peter in LA, Chanvit Lee-yaywat in Khonkaen, Hans Grosse Wilde in Germany, Marius Giphart and Jon van Rood in Leiden, Miro Radman and Dominique Charon in Paris, Wen Jie Zhang in Wuhan and many others. He was the most generous of all of us in passing on all we knew. I particularly remember how he helped Malcolm Simons before and when he was with us at RPH. Joe has yet to receive the acknowledgements he deserves. We do hope that he will find a way to continue his role as an ambassador for CYOEVF.

I am grateful to all the Board Members but I choose to single out a few.

My son, Ben has resuscitated the foundation by creating a community of occupants which, by and large, pay their leases eventually. Until recently, there has been a tendency for tenants to exploit the CYOEVF, but the Board realizes that the culture has to change if we are to contribute to the future of WA.

Roger Stubbs has inspired us with his vision and prescription for the next phase. The financial windfall of being absorbed in the urban sprawl must be devoted to renewal. We now realize this is not so much about continuing maintenance as funding the ERADE principle in new locations.

I am so pleased to have come to know Sally Lloyd, who has not only taken over responsibility for the laboratory after Joe Williamson, which was a huge achievement, but some how she seems to manage us all! I believe that, with adequate support, she can now look forward to that wonderful mix of

- creative science (with further patents for CYOEVF),
- collaborations with entrepreneurs world wide,

  and

- gentle control of present and future Board members.

Sue Lester, from the university of Adelaide, comes to participate in the quarterly retreats and just manages to satisfy the requirement that board members attend a minimum of once every 3 months. This has worked well because we are richer for the commitment, her help to Sally, Ted and me, but mainly because a periodic member brings an independent view.

Finally, I have to say that Sharon Baker has been the essential-fixed glue that keeps us functional.

Read on, because Sally and Tony Lloyd, Roger Stubbs, Sam Dawkins and Ted Steele have more to say. Also read the contributions from John Garland who, I am sure, will make our hopes and visions actually happen.

Research Director's Report

Dr Sally S Lloyd

Since joining the CYO Foundation in 2012 I have learned how valuable the haplotype approach developed here can be for gaining insight into the genetics of both human disease and animal selection. In the last year CYO Foundation's research has been increasingly focussed on using the haplotype approach to understand the genetics and evolution of cattle. The development, in 2013, of a better method of measuring the melting point of fat has allowed us to refocus this project towards selecting cattle to produce healthy beef.

Dr Joseph Williamson retired this year. The current projects at CYO Laboratory owe much to his work in developing the C19 haplotype markers for cattle. He has also passed on a wealth of knowledge on techniques to analyse DNA sequences.

Anoja Wijewickrama joined the lab part time this year, with thanks to Tillbrook Nominees for funding her position. Partially due to her contribution the lab has increased the number of C19 haplotypes tested from about 300 in 2011 to more than

1300 in 2014. This has included C19 haplotyping paternity verification for all calves born at Melaleuka Stud in 2014 and DNA tracking and fat melting point testing on all carcasses through CYO Village pub.

The increase in sample throughput has allowed us to investigate Wagyu haplotypes and their phenotypic influences. Thanks to Dom Bayard, Scott de Bruin, Keith Hammond, John Hammond and Peter Gilmore for their expertise as well as samples and data from their herds. Also thanks to Mike MacLaughlin for samples of Jersey and David Bellamy for Akaushi samples. Thanks to Ted Steele and Sue Lester for their contributions to interpreting the results.

Ted Steele and Professor Dawkins have also been responsible for keeping alive other aspects of CYO Foundation research, examining the implications of ancestral haplotypes models on individuality, evolution, artificial selection and human disease.

Research projects planned for 2015 include continuing collaboration with Scott de Bruin to investigate the heritability and haplotype associations of melting point in Wagyu, including the influence of dam haplotype compared to sire haplotypes. The first of the F1 Akaushi from Melaleuka stud will be ready for slaughter so we will be able to test the reputation of this breed for low melting point fat without requiring long grain feeding. Collaborations with Peter Thompson and CYO Village Pub will enable us to confirm the health and taste benefits of low melting point fat in beef.

Fat Melting Point

CYO's discovery of a way to use the Rotor Gene to measure the melting point of fat was published in the Journal of Animal Science in October 2014. The associate provisional patent was lodged in February 2014. We have been taking advantage of this new method to measure fat melting point on samples of Village Green beef through the CYO Pub, and Wagyu meat samples from Scott de Bruin and the Australian Wagyu Society's branded beef competition. Through these studies we are starting to recognise the influence of breed, feeding and genetics on the fat melting point of beef, the differences in melting point for different types and locations of fat within an animal and the influence of fat melting point on consumers enjoyment of beef.

One step extraction of DNA and fat

Improving the efficiency of the test process will enable it to be applied more widely. While extraction of subcutaneous fat is easily achieved by heating a small piece of tissue, extraction of intramuscular fat is more difficult. We have developed a method of extracting intramuscular fat from meat samples as part of the normal DNA extraction process. This works particularly well for meat samples with more than twenty percent fat. This has allowed us to extract DNA and fat from 0.5 gram samples taken from Wagyu carcasses at the abattoir.

Cattle Genetics

Breed C19 Haplotype Comparisons

Akaushi

Akaushi are a breed of red cattle grazed on the Kyushu island of Japan. They have a reputation for producing well marbled beef with exceptionally low melting point, without needing the expensive grain feeding used with black Wagyu. CYO has started research into the genetics and performance of Akaushi, in collaboration with Melaleuka Stud which purchased 30 full blood Akaushi cows and 15 calves in October 2013.

Eighty four Australian Akaushi were tested for C19 Haplotypes. DNA was taken from the new Akaushi herd at Melaleuka Stud. Samples were also received from Robbin Island and David Bellamy. A comparison of the C19 haplotypes of Wagyu and Akaushi show them to be entirely separate breeds, as different from each other as Wagyu are from Simmental.

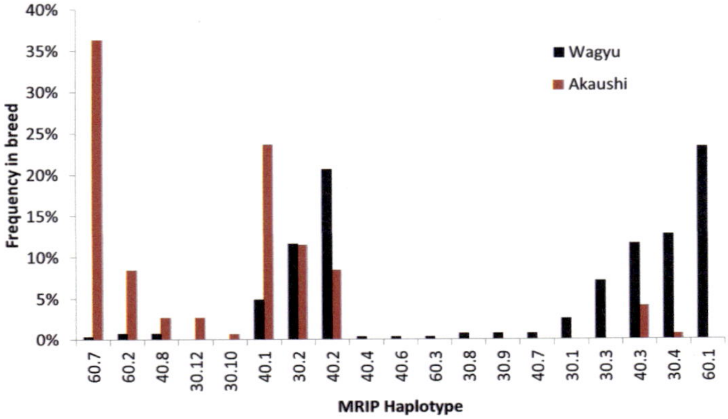

Jersey

The jersey breed is the second most popular breed of dairy cattle. It is favoured over Freisian-Holstein where there is a need for quality rather tha quantity of milk. Jersey also have a reputation for good feed efficiency, fertility and calving ease. The tendency of jersey to marble with low melting point fat is less well known, but on some measures superior even to Wagyu.

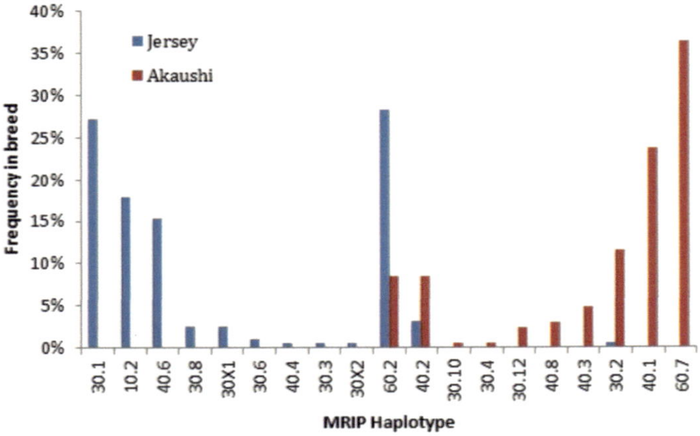

We tested the C19 haplotypes of 100 Jersey cattle from McLaughlin Dairy. This revealed a new allele of NT5M with associated new haplotypes. The most common haplotype in jersey, is only occasionally found in other breeds, but is quite common in Akaushi. It is also very similar to the most common Akaushi haplotype, 60.7.

Research Plans for 2015

In collaboration with Scott de Bruin, testing melting point and C19 haplotypes of 200 carcasses produced at Mayura station. This is along with continued samples from Peter Gilmour (Irongate Wagu, WA), Keith and John Hammond (Robbin Island, Tas) and Dom Bayard (Goorambat, Vic).

Papers Published in 2014

Lloyd, S. S., Dawkins, S. T., & Dawkins, R. L. (2014). A novel method of measuring the melting point of animal fats 1. *Journal of Animal Science*, *92*(10), 4775–4778.

Steele, E. J. (2014). Reflections on Ancestral Haplotypes: Medical Genomics, Evolution, and Human Individuality. *Perspectives in Biology and Medicine*, *57*(2), 179–197.

Lloyd, S. S., Bayard, D., Lester, S., Williamson, J. F., Steele, E. J., & Dawkins, R. L. (2014). Ancestral Haplotypes, Quantal Genomics and Healthy Beef S. In *Proceedings, 10th World Congress of Genetics Applied to Livestock Production Ancestral*.

Vice-Chairman's Report

Mr. Anthony Lloyd

Anthony Lloyd is a retired solicitor who joined the Board in 2011. He has an interest in international intellectual property law, a subject relevant to the Foundation as many of its scien- tific discoveries can be commercialised through the patent system.

He was born in the Western Australian wheat belt and initially obtained a Bachelor of Agricultural Science degree at the University of

Western Australia. Subsequently he obtained a Masters in Business Administration and a Bachelor of Laws.

Mr Lloyd has a continuing interest in the Foundation's research relevant to agricultural industries and a particular interest in applying the Foundation's research expertise to minimise frost damage to the state's grain industry.

Chromosomal Crossovers and Ancestral Haplotypes

This year is the centenary of the publication by Thomas Hunt Morgan and his colleagues of the seminal book "The Mechanism of Mendelian Hereditary".

Mendel's original work on the inheritance of certain characters in garden peas had been published in 1866 but largely overlooked until about 1900. One of the key implications of Mendel's work was that inheritance was discontinuous. It gave impetus to the search for the location and physical mechanism by which traits were inherited.

Chromosomes had been observed in the nucleus of cells since the middle of the nineteenth century. Morgan's book set out the case for the "Mendelian Factors" being physically located within the chromosomes. The book sets out Morgan's evidence for the phenomenon of "linkage", showing that different Mendelian factors tended to be inherited together ("linked"), while inheritance of others were independent. This fact supported the hypothesis that the Mendelian factors were located on the chromosomes because it was consistent with unlinked factors being located on different chromosomes and linked factors being located on the same chromosome.

Morgan's experimental work showed that "linkage" was not absolute. He was able to point to the previously observed phenomenon of chromosomal crossover during the production of reproductive cells (first reported by Janssens in 1906) as the physical manifestation of the limited recombination of Mendelian factors located on a single chromosome.

From this work, Morgan and his colleagues developed the concept of "genetic distance" – the likelihood that a crossover event would take place between two factors on a single chromosome. By measuring the "distance" between factors it was shown that factors were arranged in a stable linear sequence ("beads on a string") within a chromosome, forty years before the molecular structure of DNA was demonstrated.

The assumption made in the development of concepts of "linkage" and "genetic distance" was that chromosomal crossovers occurred more or less randomly. It was always understood that there could be variation in the rate of crossovers at different parts of a chromosome but this was a matter of probabilistic variation. "Genetic distance" might not exactly correspond to physical distance, but it was a first order approximation.

No doubt due to this thinking absorbed by generations of geneticists, there has been relatively little inquiry as to why a crossover event should occur in one place and not another. In recent years there has been work by different groups in identifying "recombination hotspots" and the molecular associations with those but even that terminology carries with it an assumption of probabilistic variation.

The work carried out at this Foundation over 30 years of genetic research has identified many regions of the genome (Ancestral Haplotypes) which must have been stable over thousands of generations. The Ancestral Haplotypes are by no means small structures, typically (150 kilobases to ≥4 megabases in length) and are functionally significant, typically encompassing many protein and regulatory coding regions. How is it that chromosomal crossovers almost never occur within such large regions? This is not a matter of probabilistic variation – the implication is that there are numerous large regions where chromosomal crossovers are entirely suppressed!

This inhibition of chromosomal crossovers within Ancestral Haplotypes is an important scientific issue. The mechanism itself may unlock our understanding of genetic processes. It also has implications for broader evolutionary issues – is relaxation of crossover inhibition a factor in the formation of the Ancestral Haplotypes (and by implication speciation events).

There are also practical animal and plant breeding implications. Breeders often wish to increase variation within a population. Lifting the mechanisms for crossover inhibition may be an efficient method of achieving "genetic progress" in domesticated plants and animals.

It is possible to postulate many molecular mechanisms by which recombination might be inhibited but they must all depend at some level on the sequences making up the Ancestral Haplotypes. The next generation of DNA sequencing should give accurate sequences for different ancestral haplotypes across the genome and enable comparison between those regions where chromosomal crossovers almost never occur and those regions where they are relatively common.

When such sequencing data becomes available, it might be possible to unravel an issue which has for far too long not received the attention it deserves.

Treasurer's Report

Mr. Wayne Cox

In 2014 our cash reserves have dropped from over the 1.0M barrier to 905k. This was caused by two major expenditures during the year. The first being some capital upgrades to the Incubators to enable us to sell them at maximum value in the coming year. Secondly, we were required to pay a redundancy for a long standing employee due to some internal restructuring of the research Laboratory.

The value of our Investment in GTG has dropped to alarming levels. The value at the end of December was 160k compared to 800k at the same time the previous year. Post the end of the year we were able to sell down 3.0M shares at an average price of 8 cents due to a sudden spike in the share price which was pleasing.

The other pleasing issue to report is our continual improvement in the rent we have been able to command with the Incubators. In the 2014 year we received $199,614 in Lease income compared to the prior year of $167,365. Thanks must go to Ben for his efforts in maximizing the Foundation's return.

Overall the Foundation made a cash loss of only 25k which is encouraging. The accounts show 1.05M loss but 940k of that loss relates to the drop in share price and the return of the Artwork.

Lastly, we have again been able to rely on the support of the AL & M Dawkins Foundation to fund our programmes.

CYO Foundation Members Profiles & Reports

Dr. Ted Steele

Ted Steele is a cellular and molecular immunologist of some 40 years standing who rejoined the CYO Foundation in 2014 after a 3 year absence. His main work in recent years has been on the mechanism of somatic hypermutation of antibody variable region genes and on rapid evolutionary mechanisms. The main theme running through all this work has been the role of RNA intermediates and reverse transcription in the generation of genetic diversity. Recently this has led to the analysis of molecular mechanisms in the early stages of cancer development. He has a deep fascination with the whole topic of ancestral haplotypes and hopes to pursue his interests around this theme at CYO.

I look back at the 2014 report and reflect on a hectic year ! – what did I *actually achieve* for the foundation? Looking at the forward plans I then outlined, several developments are now clear.

History of IRF/CYO Research Passes Peer Review

By mid 2014 it was clear that an 8,000 word article submitted to *Perspectives in Biology & Medicine* (Johns Hopkins University Press) had been accepted for publication after rather intense and testy peer review. The Title pages and Abstract are shown below. This is a synopsis history of the research achievements of CYO/IRF scientist since about 1982 in the discovery of ancestral haplotypes and establishing the quantal structure of the genome. It also flagged the significant achievements of the Boston Group of Professors Chester A Alper an Edmond J Yunis. Writing this article prepared the ground for the book length version Professor Dawkins has flagged in his Chairman's Report.

Origins of MHC Polymorphism and Alloreactivity

This project, an attempt to unify several fundamental concepts in immunology, is still ongoing. One day it may result in another book. However by mid year the germ of the ideas morphed into a surprising discovery. It involved a direct comparison of the extent of genetic diversity (polymorphism) between the ancestral haplotypes found at the MHC locus and recent similar attempts to measure the polymorphism at the IGHV locus, the site on chromosome 14, encoding the heavy chain variable genes of antibody molecules. This comparison concluded that IGHV may be many orders of magnitude *more polymorphic* than the MHC! – which hitherto has been considered the most polymorphic region in the genome.

This is a significant outcome as it strongly implies that Soma-to-Germline feedback is likely to be *commonplace* for the genes encoding antibody variable regions. i.e. somatic antibody V mutants, stimulated and selected during the course of an immune response to infectious diseases are transmitted to progeny, genetically preparing them for protection if the disease persists in the family lineage. The comparative haplotype analysis leading to this conclusion has been written up and submitted for publication. The main recommendation is that the 3-generation pedigree strategy, so successful in establishing segregation and stability of MHC ancestral haplotypes (and MRIP haplotypes on bovine chromosome 19) can now be applied to IGHV.

Cattle Breeding Program and Database

This is an ongoing involvement in the Healthy Beef program led by Research Director Dr Sally Lloyd. By April 2014 an audit of the database on Dam-Calf identification confirmed how unerring such mutual identification is at birth – we are of the view that any apparent errors in this matching maybe due to the humans doing the recording

Reflections on Ancestral Haplotypes: Medical Genomics, Evolution, and Human Individuality

Edward J. Steele

Perspectives in Biology and Medicine, Volume 57, Number 2, Spring 2014, pp. 179-197 (Article)

Published by The Johns Hopkins University Press
DOI: 10.1353/pbm.2014.0014

ABSTRACT The major histocompatibility complex (MHC), once labelled the "sphinx of immunology" by Jan Klein, provides powerful challenges to evolutionary thinking. This essay highlights the main discoveries that established the block ancestral haplotype structure of the MHC and the wider genome, focusing on the work by the Perth (Australia) group, led by Roger Dawkins, and the Boston group, led by Chester Alper and Edmond Yunis. Their achievements have been overlooked in the rush to sequence the first and subsequent drafts of the human genome. In Caucasoids, where most of the detailed work has been done, about 70% of all known allelic MHC diversity can be accounted for by 30 or so ancestral haplotypes (AHs), or conserved sequences of many mega-bases, and their recombinants. The block haplotype structure of the genome, as shown for the MHC (and other genetic regions), is a story that needs to be understood in its own right, particularly given the promotion of the "HapMap" project and single nucleotide polymorphism (SNP) linkage disequilibrium (LD) analysis, which has been wrongly touted as the only way to pinpoint those genes that are important in genetic disorders or other desired (qualitative) characteristics.

Melville Analytics Pty. Ltd., Jindabyne; and C. Y. O'Connor ERADE Village Foundation, Perth, Australia.

Correspondence: C. Y. O'Connor ERADE Village Foundation, P.O. Box 5100, Canning Vale, New South Wales 6155, Australia.

E-mail: e.j.steele@bigpond.com or ejsteele@cyo.edu.au.

The author thanks Sue Lester for providing papers difficult to access and Robyn Lindley for critical comments on the manuscript. He also acknowledges the C. Y. O'Connor ERADE Village Foundation, which supported a Fellowship during 2009–10 and provided financial assistance to attend scientific meetings. The author is currently writing a book on the discovery of ancestral haplotypes and the block haplotype structure of the genome.

The "Perth group" includes those led by Emeritus Professor R. L. Dawkins (UWA) during his past and present affiliations with the University of Western Australia, Murdoch University, and the C. Y. O'Connor ERADE Village Foundation (CYO), incorporating the Immunogenetics Research Foundation (IRF). The "Boston group" includes those led by Professors Chester A. Alper and Edmond J. Yunis at Boston Children's Hospital, Harvard University, and the Department of Pathology, Harvard Medical School.

CYO Publishing → Nearurban Publishing

The grand plan at the start of the year, was to set up an independent scientific publishing house, *CYO Publishing*. And by the end of the year it was achieved - but it too has unexpectedly morphed into a new academic publishing house *Nearurban Publishing* located in Dallas, Texas. The "NU Publishing " logo is now appended to this annual report. As with many new developments the old cliché is accurate - "necessity was the mother of the invention". Times really have changed in book publishing, particularly scientific book publishing. Established academic book publishers cater for the mainstream; they do not rock the boat. They only publish what can only be termed tracts on "dogma consolidation" (as Thomas Kuhn has taught us , this is the role of most "normal" scientific activity).

The first two books from the CYO Foundation published by *Nearurban Publishing* are described by Professor Dawkins in his report. They are now available as Paperback and Kindle versions from Amazon. The Front and Back Covers of both books are shown below. The modern Print-on-Demand (POD) publishing technology has made this all possible.

Let it not be thought we are shirking peer review – far from it. Our other published papers are running the gauntlet of peer review all the time. The importance of scholarly scientific publishing as embodied in *Nearurban* is that it allows the reflective scientist with an *innovative story* to tell, the space to develop the logic and narrative, unfettered by scientific editors who through ignorance, haste or protection can modify the message *before* it can be published. The reader should be the judge. Scientific books of this type aim to stand the test of time and they are rarely best sellers (with few notable exceptions).

Dr. Samuel Dawkins

After some years in commercial R & D, Sam has spent over a decade in academia working in the field of quantum optics, with goals including the development of the next generation of atomic clocks for probing the remaining mysteries of physics such as dark matter and unifying theories. Currently, Sam works as part of a global effort to realise a functional quantum computer, which many believe will underpin the next generation of information technology

Mr. Roger Stubbs

Roger Stubbs is a graduate of UWA and Curtin University, Roger has a background in education, town planning and consultancy.

Roger is deeply involved in Local Government; he was Mayor of Armadale for eight years and a former President of the local Government Association, also an Executive member of the Australian Local Government Association. In 2002 Roger became a Member of the Order of Australia for service to local government through affiliated professional associations to urban planning, particularly through the development of environmental management policies and to the community of Armadale.

Mr John Garland

Previously from an agricultural background John established what was then John Garland & Co in 1965 to specialize in rural associated property and business sales in Western Australia. Over the past 46 years the company's expertise has grown to one where it is now recognized as a leader in the marketing and sales of lifestyle estates, tourism ventures and the facilitation of overseas investment in Australia.

In 2007 John was awarded a Member in the General Division of the Order of Australia (AM) in recognition of his service to the Real Estate industry and the community.

A Past President of the Real Estate Institute of Western Australia (1986-88) he was recognized by that body with the Kevin Sullivan Award in 1992. He was acknowledged with a Life Membership of REIWA in 1993. John is one of Western Australia's longest practicing licensed auctioneers.

As a past Deputy World President of the Paris-based FIABCI (International Real Estate Federation), John was responsible for the Asia-Pacific area and in recognition of his contribution he is one of the very few Australians to receive the FIABCI Medal of Honour.

Dairy Industry Milk Quota Auction, designed for the Agricultural Department, provided a transition to deregulation and was a project personally designed and instigated by John. A world first! It was Modelled on the Amsterdam Flower Auction concept.

John is a past Chairman of the Small Business Development Corporation in Western Australia, a position he held for seven years. The Corporation is responsible for Western Australia's business migration centres and accepts the role of Advisory Body to the Minister for Commercial Small Business.

A Director with Future Directions International, the Major General Michael Jeffrey (Retd), Sir Charles Court organization with it's 'not-for-profit' promotion of Western Australia's future integration with Indian Ocean Rim Trade opportunities.

A Committee Member with the Regional Development Authority, a Commonwealth Government Advisory Board for allocation of Commonwealth funds to regional projects in Western Australia. He is currently a councilor with the Small Enterprise Network Advisory Body of the WA Chamber of Commerce and Industry.

John is a past Deputy Chairman of Celebrate WA, responsible for Citizen of the Year Awards, Foundation Day, Proclamation

Day, the Oz Concert and Adopt a Digger. A past committee member of the Australian British Chamber of Commerce and a Life Member of the Settlement Association of Western Australia.

The Rotary Club of Perth is another commitment, where he is a Past President and Paul Harris fellow with sapphire and is actively involved in current Rotary initiatives including being an active board member for Path of Hope which provides assistance to victims of domestic abuse through a number of initiatives, in conjunction with The Salvation Army.

John has served as a mentor for the Western Australian Police Force which included providing a weekly visit to a squad during their six months training at the Joondalup Academy.

Mr Brian Walker

Brian has been working in the finance industry for over 30 years and a great deal of that time has been involved in the Rural and Regional industry. Previous postings have included Manjimup, Esperance, Bunbury, and prior to accepting an offer within Westpac Private bank spent two years as Regional Manager overseeing the commercial and agri centres between Perth and Kununurra.

Brian's exposure to the agricultural and business industry has seen his expertise grow to where his contributions have been recognised with a number of CEO awards and also in parliamentary dispatches.

Brian has been associated with many professional bodies over the years which include The Australian Institute of Management, Australian Institute of Banking, and Chamber of Commerce and Industry.

Coupled with his financial industry service, his involvement as past President of the Chamber of Commerce and Industry (CCI) - Manjimup WA and past member of the CCI Esperance and Bunbury has given him a sound understanding of the industry needs which he hopes will give him good grounding to support the C.Y O'Connor ERADE Village Foundation.

****

IMMUNOGENETICS RESEARCH FOUNDATION WA INC
T/A
THE C.Y. O'CONNOR ERADE VILLAGE FOUNDATION
67 161 830 965

FINANCIAL REPORT
FOR THE YEAR ENDED
31st December 2014

PASCOE PARTNERS
CERTIFIED PRACTISING ACCOUNTANTS
Liability limited by a scheme approved under
Professional Standards Legislation

IMMUNOGENETICS RESEARCH FOUNDATION WA INC
T/A
THE C.Y. O'CONNOR ERADE VILLAGE FOUNDATION
67 161 830 965

CONTENTS

THE C.Y. O'CONNOR ERADE VILLAGE FOUNDATION
67 161 830 965

EMERGENCY FUND

DEPARTMENTAL TRADING, INCOME AND EXPENDITURE

FOR THE YEAR ENDED 31 DECEMBER 2014

	Note	2014 $	2013 $
EMERGENCY FUND			
OTHER INCOME		-	-
OTHER EXPENSES		-	-
SURPLUS (DEFLICIT) FROM INCOME AND EXPENDITURE		-	-
Opening Balance 1st January 2014		5,000	5,000
Closing Balance 31st December 2014		5,000	5,000

7th INTERNATIONAL MHC EVOLUTION CONFERENCE

DEPARTMENTAL TRADING, INCOME AND EXPENDITURE

FOR THE YEAR ENDED 31 DECEMBER 2014

	Note	2014 $	2013 $
7th INTERNATIONAL MHC EVOLUTION CONFERENCE			
OTHER INCOME		-	-
OTHER EXPENSES		-	-
SURPLUS (DEFLICIT) FROM INCOME AND EXPENDITURE		-	-
Opening Balance 1st January 2014		22,779	22,779
Closing Balance 31st December 2014		22,779	22,779

THE C.Y. O'CONNOR ERADE VILLAGE FOUNDATION
67 161 830 965
C Y O'CONNOR HI-TECH INCUBATOR CENTRE
DEPARTMENTAL TRADING, INCOME AND EXPENDITURE
FOR THE YEAR ENDED 31 DECEMBER 2014

	Note	2014 $	2013 $
C Y O'CONNOR HI-TECH INCUBATOR CENTRE			
OTHER INCOME			
Lease Rentals	9	199,614	167,365
EXPENSES			
Cleaning	16	440	488
Depreciation - Plant & Equipment	20	3,306	4,134
Depreciation - Buildings	21	23,565	23,169
Fees & Permits	23	-	540
Incubator Management Fees	29	141	-
Repairs & Maintenance	38	15,206	33,403
		42,658	61,733
SURPLUS (DEFICIT) FROM INCOME AND EXPENDITURE		156,957	105,632
Opening Balance 1st January 2014		1,043,698	938,066
Closing Balance 31st December 2014		1,200,655	1,043,698

THE C.Y. O'CONNOR ERADE VILLAGE FOUNDATION

67 161 830 965

RESEARCH & GENERAL FUND

DEPARTMENTAL TRADING, INCOME AND EXPENDITURE

FOR THE YEAR ENDED 31 DECEMBER 2014

	Note	2014 $	2013 $
RESEARCH & GENERAL			
Other Income	2	280	-
Donations/Bequests	3	131,411	138,726
Fund Raising	4	-	7,050
Meeting Registration Fees	5	7,130	13,927
Membership Fees	6	8,318	-
Laundry & Gaming Machine sales	7	2,593	2,878
		149,732	162,581
OTHER INCOME			
Interest Received	8	32,395	45,195
Reimbursements & Recoups	10	11,393	9,740
Profit on Sale of Non-current Assets	11	110	5,368
		43,898	60,304
EXPENSES			
Accountancy & Audit Fees	12	7,850	-
Advertising	13	80	6,640
Art Works Returned	14	299,658	-
Bank Charges	15	180	298
Cleaning	16	2,556	5,975
Consultancy Fees	17	7,819	12,630
Contract Work	18	-	15,000
Computer Expense	19	480	-
Depreciation - Plant & Equipment	20	23,266	27,508
Bad Debts Written Off	22	-	2,513
Fees & Permits	23	1,594	2,380
Freight & Cartage	24	323	675
Fund Raising Costs	25	30	-
Insurance	26	10,534	11,073
Legal Costs	27	518	-
Licensing Fees	28	-	680
Meeting/Visitors/Honoraria	30	12,743	3,932
Consumables & Lab Supplies	31	30,222	31,364
Motor Vehicle Expense	32	772	-
Patent Fees	33	-	15,529
Prior Year Adjustments	34	-	(11,276)
Printing & Stationery	35	10,433	10,352

Fundraising Expenses	36	-	2,000
Rent	37	-	581
Repairs & Maintenance	38	5,284	383
Scholarships	39	30,000	1,000
Subscriptions	40	-	522
Superannuation Contributions	41	20,692	17,153
Telephone	42	11,453	9,857
Travelling Expenses	43	2,008	-
Unrealised Loss on Shares	44	640,667	320,333
Wages	45	245,651	162,346
Wages Tax	45	-	48,818
Waste Disposal	46	1,159	611
		1,365,969	698,878
SURPLUS (DEFLICIT) FROM INCOME AND EXPENDITURE		(1,172,339)	(475,993)
Opening Balance 1st January 2014		2,126,649	2,602,643
Closing Balance 31st December 2014		954,310	2,126,649

THE C.Y. O'CONNOR ERADE VILLAGE FOUNDATION

67 161 830 965

BALANCE SHEET

AS AT 31 DECEMBER 2014

	Note	2014 $	2013 $
TRUST FUNDS			
RESERVES	47	2,182,744	3,198,126
TOTAL TRUST FUNDS		2,182,744	3,198,126
Represented by:			
CURRENT ASSETS			
Cash on Hand		1	1
Petty Cash		183	91
Deposits on Building Contracts		72,439	-
Westpac Max 112867		39,510	3,785
Westpac Research Account 10-4752		1,872	-
Westpac Community Solutions Account 1645		14,175	4,743
Accounts Receivable		26,951	29,544
ATO Account		-	2,071
Tax clearing account		-	667
Listed Shares in Public Companies	48	160,167	800,833
TOTAL CURRENT ASSETS		315,298	841,734
NON CURRENT ASSETS			
Fixed Assets			
Computer Equipment		364,971	364,971
Less: Accumulated Depreciation		(361,946)	(360,029)
		3,025	4,942
Motor Vehicles		13,500	13,500
Less: Accumulated Depreciation		(12,303)	(11,955)
		1,197	1,545
Office Furniture & Equipment		11,612	11,612
Less: Accumulated Depreciation		(10,774)	(10,565)
		838	1,047
Laboratory Equipment		507,296	507,296
Less: Accumulated Depreciation		(406,845)	(386,053)
		100,451	121,243
Incubators		51,392	51,392

Less: Accumulated Depreciation		(38,164)	(34,858)
		13,228	16,534
Incubator Buildings		1,105,241	1,105,241
Less: Accumulated Depreciation		(234,531)	(210,966)
		870,709	894,274
Total Fixed Assets		989,448	1,039,585
Investments			
Non Financial Investments	49	-	299,658
Patents	50	54,750	13,785
Term Deposits	51	850,000	1,031,084
Total Investments		904,750	1,344,527
TOTAL NON CURRENT ASSETS		1,894,198	2,384,113
TOTAL ASSETS		2,209,496	3,225,847
CURRENT LIABILITIES			
Westpac Research Account 10-4752		-	5,633
Trade Creditors		9,033	7,481
Incubator Bond Monies	52	11,810	8,720
Amounts Withheld	53	4,649	5,887
Tax clearing account		1,260	-
TOTAL CURRENT LIABILITIES		26,752	27,721
TOTAL LIABILITIES		26,752	27,721
NET ASSETS		2,182,744	3,198,126

THE C.Y. O'CONNOR ERADE VILLAGE FOUNDATION
67 161 830 965

NOTES TO THE FINANCIAL STATEMENTS

FOR THE YEAR ENDED 31 DECEMBER 2014

	2014	2013
	$	$

1 Statement of Significant Accounting Policies

The director of the trustee company has prepared the financial statements of the trust on the basis that the trust is a non-reporting entity because there are no users dependent on general purpose financial statements. The financial statements are therefore special purpose financial statements that have been prepared in order to meet the needs of beneficiaries.

The financial statements have been prepared in accordance with the significant accounting policies disclosed below which the director has determined are appropriate to meet the needs of members. Such accounting policies are consistent with the previous period unless stated otherwise.

The financial statements have been prepared on an accruals basis and are based on historical costs unless otherwise stated in the notes. The accounting policies that have been adopted in the preparation of these statements are as follows:

Property, Plant and Equipment

Property, plant and equipment are carried at cost, independent or director's valuation. All assets, excluding freehold land and buildings are depreciated over the useful lives of the assets to the trust.

Increases in the carrying amount arising on revaluation of land and buildings are credited to a revaluation reserve in shareholders' equity. Decreases that offset previous increases of the same asset are charged against fair value reserves directly in equity; all other decreases are charged to the income statement. Each year the difference between depreciation based on the revalued carrying amount of the asset charged to the income statement and depreciation based on the asset's original cost is transferred from the revaluation reserve to retained earnings.

The carrying amount of plant and equipment is reviewed annually by directors to ensure it is not in excess of the recoverable amount from these assets. The recoverable amount is assessed on the basis of the expected net cash flows that will be received from the asset's employment and subsequent disposal. The expected net cash flows have not been discounted in determining recoverable amounts.

Property, Plant and Equipment (PPE)

Leasehold improvements and office equipment are carried at cost less, where applicable, any accumulated depreciation.

The depreciable amount of all PPE is depreciated over the useful lives of the assets to the trust commencing from the time the asset is held ready for use.

Leasehold improvements are amortised over the shorter of either the unexpired period of the lease or the estimated useful lives of the improvements.

THE C.Y. O'CONNOR ERADE VILLAGE FOUNDATION
67 161 830 965

NOTES TO THE FINANCIAL STATEMENTS
FOR THE YEAR ENDED 31 DECEMBER 2014

		2014 $	2013 $
2	**Other Income**		
	Other Income	280	-
		280	-
3	**Donations/Bequests**		
	Donations/Bequests	131,411	138,726
		131,411	138,726
4	**Fund Raising**		
	Fund Raising	-	7,050
		-	7,050
5	**Meeting Registration Fees**		
	Meeting Registration Fees	7,130	13,927
		7,130	13,927
6	**Membership Fees**		
	Membership Fees	8,318	-
		8,318	-
7	**Laundry & Gaming Machine sales**		
	Laundry & Gaming Machine sales	2,593	2,878
		2,593	2,878
8	**Interest Received**		
	Interest Received	32,395	45,195
		32,395	45,195
9	**Lease Rentals**		
	Lease Rentals	199,614	167,365
		199,614	167,365
10	**Reimbursements & Recoups**		
	Reimbursements & Recoups	11,393	9,740
		11,393	9,740
11	**Profit on Sale of Non-current Assets**		
	Profit on Sale of Non-current Assets	110	5,368
		110	5,368
12	**Accountancy & Audit Fees**		
	Accountancy & Audit Fees	7,850	-
		7,850	-
13	**Advertising**		
	Advertising	80	6,640

		80	6,640
14	**Art Works Returned**		
	Art Works Returned	299,658	-
		299,658	-
15	**Bank Charges**		
	Bank Charges	180	298
		180	298
16	**Cleaning**		
	Cleaning	440	488
	Cleaning	2,556	5,975
		2,996	6,463
17	**Consultancy Fees**		
	Consultancy Fees	7,819	12,630
		7,819	12,630
18	**Contract Work**		
	Contract Work	-	15,000
		-	15,000
19	**Computer Expense**		
	Computer Expense	480	-
		480	-
20	**Depreciation - Plant & Equipment**		
	Depreciation - Plant & Equipment	3,306	4,134
	Depreciation - Plant & Equipment	23,266	27,508
		26,572	31,642
21	**Depreciation - Buildings**		
	Depreciation - Buildings	23,565	23,169
		23,565	23,169
22	**Bad Debts Written Off**		
	Bad Debts Written Off	-	2,513
		-	2,513
23	**Fees & Permits**		
	Fees & Permits	-	540
	Fees & Permits	1,594	2,380
		1,594	2,920
24	**Freight & Cartage**		
	Freight & Cartage	323	675
		323	675
25	**Fund Raising Costs**		
	Fund Raising Costs	30	-
		30	-
26	**Insurance**		

	Insurance	10,534	11,073
		10,534	11,073
27	**Legal Costs**		
	Legal Costs	518	-
		518	-
28	**Licensing Fees**		
	Licensing Fees	-	680
		-	680
29	**Incubator Management Fees**		
	Incubator Management Fees	141	-
		141	-
30	**Meeting/Visitors/Honoraria**		
	Meeting/Visitors/Honoraria	12,743	3,932
		12,743	3,932
31	**Consumables & Lab Supplies**		
	Consumables & Lab Supplies	30,222	31,364
		30,222	31,364
32	**Motor Vehicle Expenses**		
	Motor Vehicle Expense	772	-
		772	-
33	**Patent Fees**		
	Patent Fees	-	15,529
		-	15,529
34	**Prior Year Adjustments**		
	Prior Year Adjustments	-	(11,276)
		-	(11,276)
35	**Printing & Stationery**		
	Printing & Stationery	10,433	10,352
		10,433	10,352
36	**Fundraising Expenses**		
	Fundraising Expenses	-	2,000
		-	2,000
37	**Rent**		
	Rent	-	581
		-	581
38	**Repairs & Maintenance**		
	Repairs & Maintenance	15,206	33,403
	Repairs & Maintenance	5,284	383
		20,489	33,785
39	**Scholarships**		
	Scholarships	30,000	1,000

		30,000	1,000
40	**Subscriptions**		
	Subscriptions	-	522
		-	522
41	**Superannuation Contributions**		
	Superannuation Contributions	20,692	17,153
		20,692	17,153
42	**Telephone**		
	Telephone	11,453	9,857
		11,453	9,857
43	**Travelling Expenses**		
	Travelling Expenses	2,008	-
		2,008	-
44	**Unrealised Loss on Shares**		
	Unrealised Loss on Shares	640,667	320,333
		640,667	320,333
45	**Wages**		
	Wages	245,651	162,346
	Wages Tax	-	48,818
		245,651	211,164
46	**Waste Disposal**		
	Waste Disposal	1,159	611
		1,159	611
47	**RESERVES**		
	Emergency Fund	5,000	5,000
	7th MHC Evolution Meeting Fund	22,779	22,779
	C Y O'Connor Hi-Tech Incubator Centre	1,200,655	1,043,698
	Research & General Fund	954,310	2,126,649
	Reserves - Surplus (Deficit)	-	-
		2,182,744	3,198,126
48	**Listed Shares in Public Companies**		
	Shares in Listed Companies	12,012,500	12,012,500
	Less Provision for Diminution in Value	(11,852,333)	(11,211,667)
		160,167	800,833

Listed Shares in Public Companies include 16,016,667 shares held in Genetic Technologies Ltd. In 2004 the Foundation sold patents to Genetic Technologies Ltd. The agreement provided for the sale of the patents by the Foundation to Genetic Technologies Ltd for an amount of $15,000,000 and the granting of a licence by Genetic Technologies Ltd the foundation for the use of patents, at a cost of $2,500,000. The sale agreement provided the Foundation would receive net consideration of $12,500,000 to be Satisfied by 16,666,667 shares at an agreed price of 75 cents each. The shares were issued to the The Foundation on the 16th of June 2004. The shares were subject to escrow. Twenty percent of the total shareholding came out of escrow on each anniversary of the acquisition date. The ASX share price of Genetic Technologies Ltd at 31 December 2014 was 1 cents per share. Although the shareholding is considered a long term investment by the foundation, in accordance with normal accounting practice, the Foundation has provided $11,852,333 for the diminution in the value of the shares between the issue price and the market price at 31 December 2014.

49	**Non Financial Investments**		
	Art Works	-	299,658
		-	299,658
50	**Patents**		
	Patents	54,750	13,785
		54,750	13,785
51	**Term Deposits**		
	Westpac Term Deposit 17-2974	-	261,250
	Westpac Term Deposit 16-3971	-	265,436
	Westpac Term Deposit 17-7599	125,000	254,398
	Westpac Term Deposit 18-6946	125,000	125,000
	Westpac Term Deposit 21-4134	250,000	125,000
	Westpac Term Deposit 21-4126	125,000	-
	Westpac Term Deposit 21-5671	125,000	-
	Westpac Term Deposit 21-9453	100,000	-
		850,000	1,031,084
52	**Incubator Bond Monies**		
	Auslanka	1,120	1,120
	Councilling Connectior	-	1,120
	LimeLight	720	720
	National Security	-	1,120
	Better Care Medical Centre	1,360	1,360
	Youme Clothing	450	450
	Spicy Wok	1,200	1,200
	Make Your Mark Art Cafe	1,040	1,040
	Infinity Wealth Solutions	320	320
	Luke Nixon	-	270
	Renov8withus Pty Ltd	1,400	-
	Indebiti Solutions Pty Ltd	1,400	-
	Alight Scent	1,400	-
	Mr Stubby	1,400	-
		11,810	8,720

53	**Amounts Withheld**		
	Amounts Withheld from Salary, Wages and Other Payments	1,284	4,523
	Employee Superannuation Payable	3,365	1,364
		4,649	5,887

COMPILATION REPORT
THE C.Y. O'CONNOR ERADE VILLAGE FOUNDATION
67 161 830 965

We have compiled the accompanying special purpose financial statements of THE C.Y. O'CONNOR ERADE VILLAGE FOUNDATION which comprise the balance sheet as at 31 December 2014, profit and loss statement and cash flow statement for the year then ended, a summary of significant accounting policies and other explanatory notes.

The specific purpose for which these special purpose financial statements have been prepared is to provide informatior

The responsibility of the Director

The director is solely responsible for the information contained in the special purpose financial statements and has determined that the basis of accounting used is appropriate to meet their needs and for the purpose for which the financial statements were prepared.

Our responsibility

On the basis of information provided by the director, we have compiled the accompanying special purpose financial statements in accordance with the financial reporting framework described in Note 1 to the financial statements and APES 315: Compilation of Financial Information.

Our procedures use accounting expertise to collect, classify and summarise the financial information, which the director provided, into a financial report. Our procedures do not include verification or validation procedures. No audit or review has been performed and accordingly no assurance is expressed.

The special purpose financial statements were prepared exclusively for the director. We do not accept responsibility to any other person for the content of the special purpose financial statements.

Name of Firm:	Pascoe Partners Certified Practising Accountants
Name of Partner:	Wayne Michael Cox *MTax, Dip FS(FP), CPA, CTA*
Address:	Level 2, 100 Havelock Street West Perth
Dated:	11 · 3 · 15

The C.Y. O'Connor Erade Village Foundation
Independent Auditor's Report
to the Members of
The C.Y. O'Connor Erade Village Foundation
Incorporating
The Immunogenetics Research Foundation
and
The Institute of Molecular Genetics and Immunology Incorporated

---

We have audited the accompanying financial report, being a special purpose financial report, of The C.Y. O'Connor Erade Village Foundation (the association), which comprises the committee's report, the assets and liabilities statement as at 31 December 2014, the income and expenditure statement for the year then ended, notes comprising a summary of significant accounting policies and other explanatory information, and the statement by members of the committee.

Committee's Responsibility for the Financial Report

The committee of The C.Y. O'Connor Erade Village Foundation is responsible for the preparation of the financial report, and has determined that the basis of preparation described in Note 1 is appropriate to meet the requirements of the Associations Incorporation Act (1987) and is appropriate to meet the needs of the members. The committee's responsibility also includes such internal control as the committee determines is necessary to enable the preparation of a financial report that is free from material misstatement, whether due to fraud or error.

Auditor's Responsibility

Our responsibility is to express an opinion on the financial report based on our audit. We have conducted our audit in accordance with Australian Auditing Standards. Those standards require that we comply with relevant ethical requirements relating to audit engagements and plan and perform the audit to obtain reasonable assurance whether the financial report is free from material misstatement.

An audit involves performing procedures to obtain audit evidence about the amounts and disclosures in the financial report. The procedures selected depend on the auditor's judgment, including the assessment of the risks of material misstatement of the financial report, whether due to fraud or error. In making those risk assessments, the auditor considers internal control relevant to the association's preparation of the financial report that gives a true and fair view, in order to design audit procedures that are appropriate in the circumstances, but not for the purpose of expressing an opinion on the effectiveness of the association's internal control. An audit also includes evaluating the appropriateness of accounting policies used and the reasonableness of accounting estimates made by the committee, as well as evaluating the overall presentation of the financial report.

We believe that the audit evidence we have obtained is sufficient and appropriate to provide a basis for our audit opinion.

Qualification

In common with most charitable institutions receiving cash donations, the entity has determined that it is impractical to establish controls over cash donations before they are banked and recorded in the accounting system. As a result our audit was restricted to the amounts recorded in the accounting system.

Opinion

In our opinion, except for the effects on the financial report, if any that the above qualification may have on the financial report of The C.Y. O'Connor Erade Village Foundation, the report presents fairly, in all material respects, the financial position of The C.Y. O'Connor Erade Village Foundation as at 31 December 2014 and its financial performance for the year then ended in accordance with the accounting policies described in Note 1 to the financial statements, and the Associations Incorporation Act (1987).

The C.Y. O'Connor Erade Village Foundation
Independent Auditor's Report
to the Members of
The C.Y. O'Connor Erade Village Foundation
Incorporating
The Immunogenetics Research Foundation
and
The Institute of Molecular Genetics and Immunology Incorporated

---

Basis of Accounting and Restriction on Distribution
Without modifying our opinion, we draw attention to Note 1 to the financial report, which describes the basis of accounting. The financial report has been prepared to assist The C.Y. O'Connor Erade Village Foundation to meet the requirements of the Associations Incorporation Act (1987). As a result, the financial report may not be suitable for another purpose.

S. Williams

Steve Williams FIPA
Auditor

Dated this..2nd..day of....APRIL....2015

Auditors Independence Declaration

Under Section 307C of the Corporations Act 2001

To Members of The C.Y. O'Connor Erade Village Foundation

I declare that, to the best of my knowledge and belief, in relation to the audit of The C.Y. O'Connor Erade Village Foundation for the year ended 31 December 2014 there have been:

a) no contraventions of the auditor independence requirements of the Corporations Act 2001 in relation to the audit; and

b) no contraventions of any applicable code of professional conduct in relation to the audit.

S. Williams

Steve Williams FIPA
Auditor

Dated this..2ND..day of....APRIL....2015

Recent Foundation Publications

Book Covers - Nearurban Publications

Front

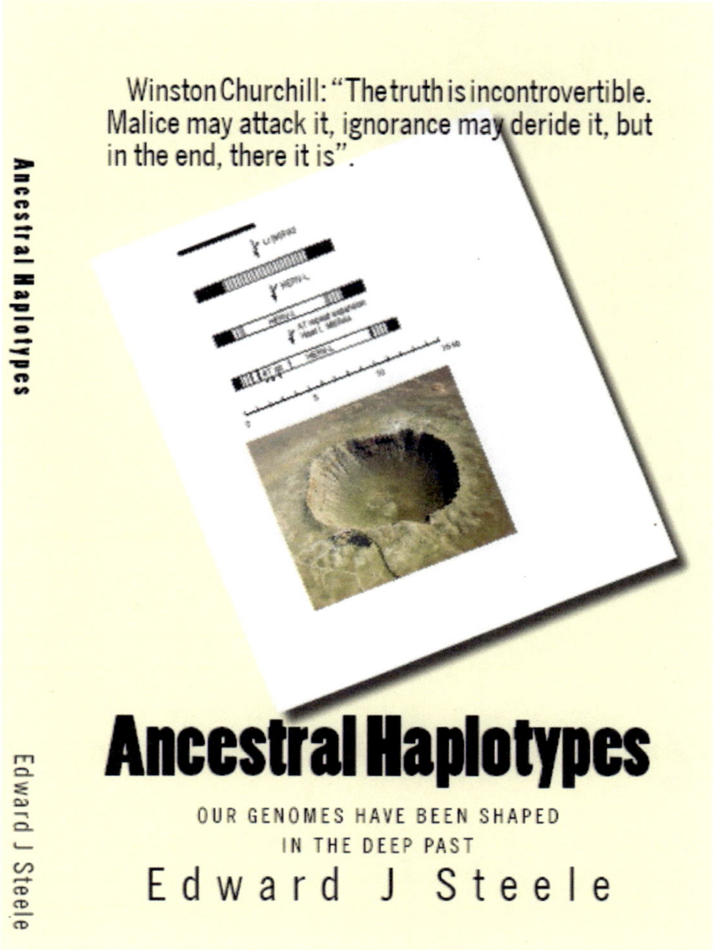

Back

This book recounts a great scientific discovery. It tells the story of the leviathan effort to demonstrate that the structure of the DNA sequences in our chromosomes is not what it might seem. A naive view of evolution has invaded mainstream scientific thinking: that all our important genetic inheritance can be viewed under a simple Darwinian view of evolution based on recent and ongoing random gene mutation followed by 'natural selection'. To be more explicit, the neo-Darwinian paradigm based on Population Genetics and statistics. The discovery of "ancestral haplotypes" radically changes this simple genetic view. In most regions of the genome many of our important genes are very old, dating to mankind's migration out of Africa ca. 100,000 years ago. They occur in large segmentally duplicated clusters - chromosomal blocks which carry tens to hundreds of genes "frozen in time". These conserved blocks are shuffled at each generation to produce individual human beings, who, for the most part, are healthy and robust and well adapted. These are our "ancestral haplotype" sequences and they are the dominant feature of genomic structure, not the simple "beads on a string" model of traditional Population Genetics and neo-Darwinism. This book tells the scientific story of their discovery and the new picture of the genome it reveals.

NU Publishing

Barcode Area

We will add the barcode for you.

Made with Cover Creator

Front

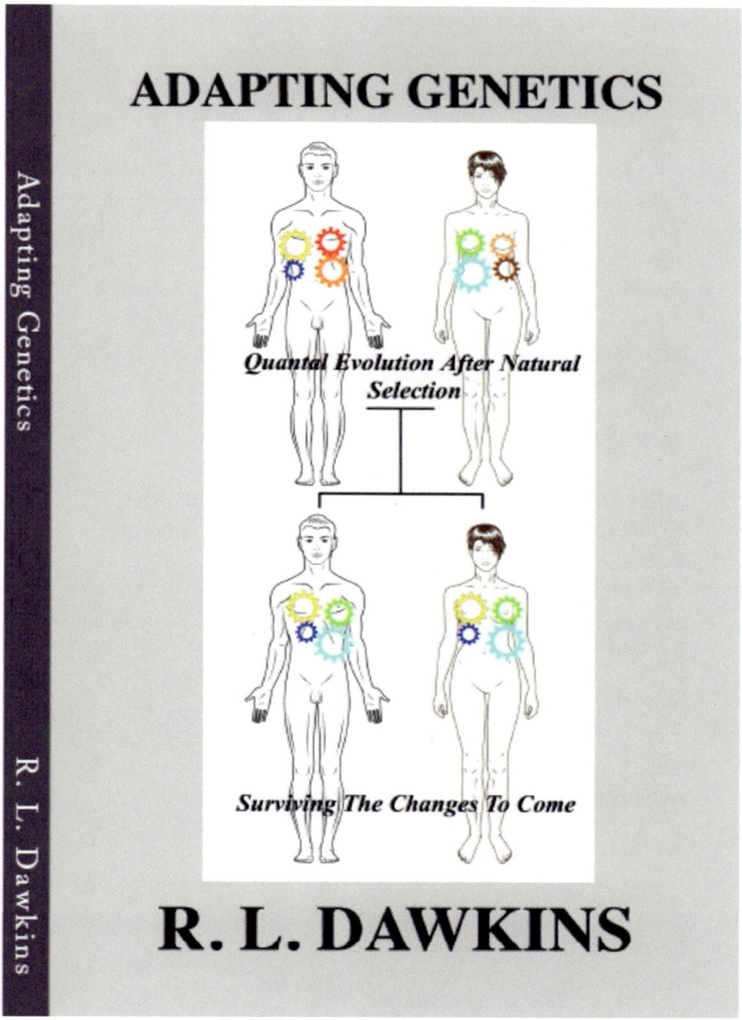
ADAPTING GENETICS
Quantal Evolution After Natural Selection
Surviving The Changes To Come
R. L. DAWKINS
Adapting Genetics
R. L. Dawkins

Back

Genetics and evolution, as taught for a century, imply that survival is a lottery dependent upon random errors in copying the genome.

Quantal evolution refers to the partitioning of the genome to protect critical regulatory DNA sequences, including during the successive ***retroviral*** epidemics which defined human ***ancestors***. Trial and error selected pre-tested combinations which are faithfully inherited thereby allowing the new species to adapt by changing gears in the ***absence*** of natural selection or mutations. Heterozygous ***polymorphisms*** interact polymorphically, thereby creating additional options. There are clear implications for ***forensics, livestock selection*** and ***transplantation*** inter alia.

Anamnestic evolution ensures individuality, diversity, protection from repeat epidemics and survival in the future.

" I enjoyed the book very much. It is a refreshing look at genetic evolution, which is really needed.
.....Professor Stephen Oppenheimer, author of "Out of Eden" and other books

" I could not stop reading it. How good it is to see that Professor Dawkins has not changed his enjoyable way of arguing."
....Professor Jon van Rood, leading HLA pioneer who's work helped establish the clinical and scientific study of the Major Histocompatibility Complex.

NU
Publishing

Barcode Area

We will add the barcode for you.

Made with Cover Creator

A novel method of measuring the melting point of animal fats[1]

S. S. Lloyd,[2] S. T. Dawkins, and R. L. Dawkins

CY O'Connor ERADE Village Foundation, 11 Erade Drive, Piara Waters, Western Australia 6112

ABSTRACT: The melting point (T_M) of fat is relevant to health, but available methods of determining T_M are cumbersome. One of the standard methods of measuring T_M for animal and vegetable fats is the slip point, also known as the open capillary method. This method is imprecise and not amenable to automation or mass testing. We have developed a technique for measuring T_M of animal fat using the Rotor-Gene Q (Qiagen, Hilden, Germany). The assay has an intra-assay SD of 0.08°C. A single operator can extract and assay up to 250 samples of animal fat in 24 h, including the time to extract the fat from the adipose tissue. This technique will improve the quality of research into genetic and environmental contributions to fat composition of meat.

Key words: animal fat, beef, fatty acid, melting point

 J. Anim. Sci. 2014.92:4775–4778
doi:10.2527/jas2014-7981

INTRODUCTION

The type of fat in the diet is as important as the quantity for maintaining good health. Diets that replace SFA with MUFA or PUFA are similar to or better than diets that replace SFA with carbohydrates. These effects can be seen in fasting glucose (Paniagua et al., 2007) and cholesterol profile (Mensink et al., 2003) and on coronary heart disease (Micha and Mozaffarian, 2010). The effect of MUFA and PUFA content on palatability and health is associated with their lower melting points (**T_M**). Even within the SFA, lauric acid has a lower T_M than stearic or palmitic acid (Knothe and Dunn, 2009) and a better effect on the ratio of total cholesterol to high-density lipoprotein cholesterol (Mensink et al., 2003).

Blood lipids improve when beef with high SFA content is replaced in the diet by beef with high MUFA content (Adams et al., 2010; Gilmore et al., 2011). Increased MUFA content also improves the flavor and texture of beef (Garmyn et al., 2011). Research into reducing the T_M of beef fat and increasing its MUFA and PUFA content is occurring (Pitchford et al., 2002; De Smet et al., 2004; Daley et al., 2010; Uemoto et al., 2010) but would benefit from a better method of measuring T_M.

Currently there are several fat T_M methods, of which the most commonly used is the slip point or open capillary method (American Oil Chemists' Society official method Cc 3b-92; AOCS, 1998). Other methods are less commonly used because they are more cumbersome, expensive, or inaccurate. All these methods lack the capacity for real-time multiple sample testing.

Therefore, an accurate T_M assay suitable for testing fat from large numbers of animals is desirable. Here we describe a method, using the Rotor-Gene Q real-time PCR Cycler (Qiagen, Hilden, Germany), which allows T_M measurements on up to 71 fat samples per 20-min run.

MATERIALS AND METHODS

Rotor-Gene Q Assay

The Rotor-Gene Q features rapid and precise temperature control, centrifugation, and real-time fluorescence detection that, although developed for high-resolution melting analysis of DNA, are exactly the components required to perform a unique measurement of fat T_M.

Twenty microliters of melted fat samples were transferred into 0.1-mL tubes. The tubes were then centrifuged at 9.8 × g and at 75°C for 2 min to collect the fat at the base of the tubes. The fat was solidified at 4°C for at least 12 h. Then, 10 μL of 26 *μM* fluorescein dye in deionized distilled water was added to form a layer above the fat. The samples were conditioned at 25°C for 90 s in the Rotor-Gene Q assembly. During

[1]The research was supported by the CY O'Connor ERADE Village Foundation. The authors would like to thank Scott de Bruin, Dom Bayard, and J.R. Dawkins for fat samples from their cattle.
[2]Corresponding author: slloyd@cyo.edu.au
Received April 29, 2014.
Accepted July 22, 2014.

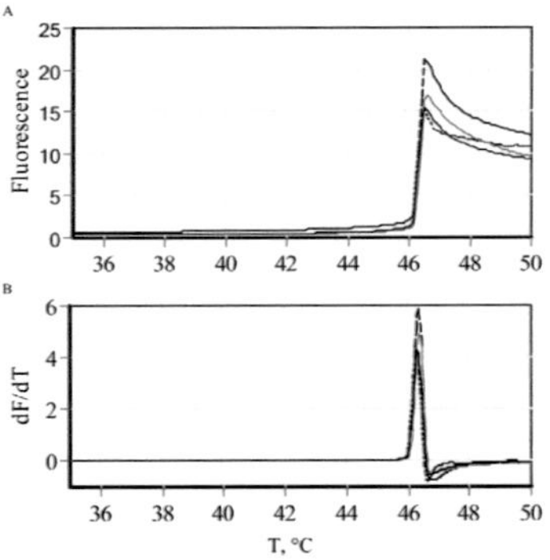

Figure 1. Fluorescence–temperature curves for 4 aliquots of rendered animal fat. Fluorescence is measured in relative units as a percentage of the maximum (saturated) signal. The derivative of fluorescence with respect to temperature (dF/dT) was calculated from the smoothed data using the Rotor Gene Q software.

the conditioning step, the photomultiplier gain was optimized using a tube loaded with dye only, which provided a reference fluorescence level.

Fluorescence detection in the Rotor-Gene Q is focused on the base of each tube. At the start of a run, the dye was above the fat and the fluorescence detection was low (see Fig. 1a for temperatures below 45°C). During the run, the temperature increased in 0.1°C steps with a rest time of 2 s at each step. At the T_M transition there was a sharp increase in fluorescence (Fig. 1a; at about 46°C). Gravitational and centrifugal forces pulled the dye down, through the liquefied fat, into the field of view of the detector, increasing the fluorescence signal. The inversion occurred because fat is less dense than the aqueous dye.

For convenience, and to ease automation, we define the T_M as the temperature at which the fluorescence signal changes most rapidly. This temperature, at the peak in the derivative of fluorescence with respect to temperature (shown in Fig 1b) is calculated automatically by the Rotor-Gene Q software.

Samples

We chose 3 standards within the T_M range of interest:

1. A sample of adipose tissue taken from the skirt fat of a Simmental steer slaughtered at 12 mo. We heated the adipose tissue to 90°C for 12 h and separated the melted fat from the water and solids by pipette.
2. Supafry (Goodman Fielder, NSW, Australia) is a commercially available rendered animal fat with a saturated fat content of 48% (wt/wt).
3. Allowrie Ghee (Fonterra, VIC, Australia) is clarified butter and has a saturated fat content of 66% (wt/wt).

Aliquots of each standard were stored at 4°C.

Fat samples were taken from 67 full-blood Wagyu steer carcasses slaughtered at an average age of 26.3 mo (SD = 1.7) after an average of 320 d on feed (SD = 14). Approximately 1 mL of adipose tissue was scraped from the subcutaneous fat layer between the 10th and 11th rib of each carcass.

To provide a comparison we sampled the subcutaneous fat from the sirloins of 17 Simmental carcasses. These were slaughtered at between 9 and 14 mo, after 36 to 87 d on feed.

We heated the carcass samples, using the method described above, to extract melted fat for T_M measurements.

Slip Point Measurements

We measured slip point temperature for each standard. First, melted fat was drawn into a glass capillary tube with an internal diameter of 1 mm to form a 1-cm column. The fat solidified in the capillary tubes while stored at 4°C for at least 12 h. The position of the top of the fat column was marked on the capillary tubes, which were then attached to a thermometer marked in 1°C divisions. The thermometer was held in a beaker of water, which was heated gradually from 20 to 50°C at 1°C/min with the top of the fat column below the surface of the water. The temperature at which the fat column rose was recorded as the slip point. We measured slip points 4 times for each fat standard.

RESULTS AND DISCUSSION

Table 1 shows the Rotor-Gene Q T_M measurements of 7 aliquots of 3 different types of animal fat over 2 runs. Variation in T_M measurements on duplicated samples within a run was very low. We found that the variation between duplicates was small both within a run (average intra-assay SD of 0.08°C) and between runs (average interassay SD of 0.10°C). The assay was accurate over the range 35 to 47°C.

Currently, there are several fat T_M methods used depending on the purpose of the test. These include the slip point or open capillary method (American Oil Chemists' Society official method Cc 3b-92; AOCS, 1998), the softening point (American Oil Chemists' Society official method Cc 3-25; AOCS, 1998), the point at which the fat becomes clear (American Oil Chemists' Society official method Cc 1-25; AOCS, 1998), the dropping point (American Oil Chemists' Society official method Cc

Table 1. High-precision melting point (T_M) measurements for 3 animal fats

Sample number	T_M, °C		
	Ghee[1]	Skirt fat	Supafry[2]
Run 1			
Aliquot 1	35.42	40.90	46.32
Aliquot 2	35.40	41.07	46.25
Aliquot 3	35.43	41.17	46.35
Aliquot 4	35.52	41.40	46.28
Mean	35.44	41.14	46.30
SD	0.05	0.21	0.04
Run 2			
Aliquot 1	35.43	41.27	46.37
Aliquot 2	35.43	41.35	46.42
Aliquot 3	35.52	41.40	46.33
Mean	35.46	41.34	46.37
SD	0.05	0.07	0.05

[1]Allowrie Ghee (Fonterra, VIC, Australia).

[2]Goodman Fielder, NSW, Australia.

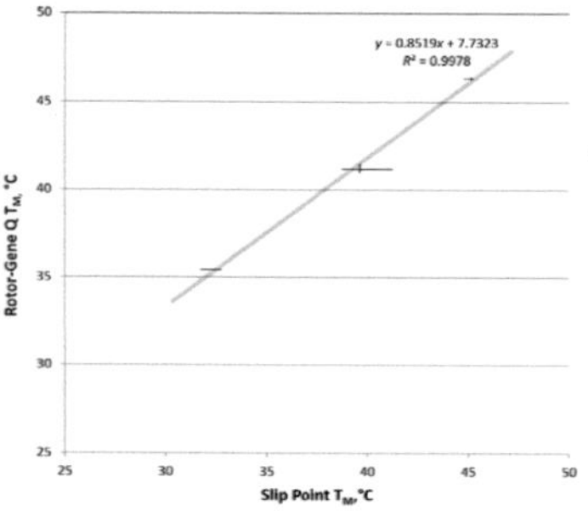

Figure 2. Correlation of melting point (T_M) measurements by slip point and Rotor-Gene Q (Qiagen, Hilden, Germany) methods.

18-80; AOCS, 1998), and differential scanning calorimetry (AOCS, 1998; Nassu and Gonçalves, 1999). Since animal fats are mixtures, their melting is a gradual process and the different definitions of the T_M in the various methods make it difficult to compare results (Nassu and Gonçalves, 1999). The slip point is the most commonly used T_M measure for fats in food production.

In Fig. 2 we compare T_M measurements using the Rotor-Gene Q assay with those using the slip point assay. The results are highly correlated but the slip point assay had more variability than the Rotor-Gene Q assay. Error bars show the large range of repeated measurements for the 4 slip point measurements compared to the small range of the 7 Rotor-Gene Q measurements. The Rotor-Gene Q T_M assay can be used in place of the standard slip point assay for various animal fats. Linear conversion is possible with Rotor-Gene Q measurements 3.3°C higher for slip points of 30 and 1.1°C higher for slip points of 45°C.

The slip point assay depends on observing fat solidified in a capillary tube rising when it melts. Slip point testing can be automated, to some extent, with purpose-designed machines. These improve the precision but not the throughput. The Melting Point M-565 (Buchi Laboratory Equipment, Flawil, Switzerland) can be used for slip point measurements of up to 3 samples at a time with a precision of 0.1°C. The Rotor-Gene Q assay described here is more reliable and accurate and is capable of testing up to 71 samples at a time.

As an example of potential applications of this technique, we measured T_M of subcutaneous fat from 67 Wagyu carcasses and 17 Simmental carcasses. We sampled 1 carcass twice, to give an indication of sampling variability, and the T_M measurements differed by less than 0.2°C. The T_M of the Wagyu samples ranged from 32.5 to 39.5°C with an average of 36.0°C and SD of 1.3°C (Fig. 3). In contrast, the Simmental samples ranged from 39.6 to 45.1°C with an average of 42.1°C and SD of 1.7°C. The phase of a fatty acid at body temperature (37°C) is known to be critical to its palatability and health effects. Of the 66 Wagyu carcasses sampled, 52 had fat that was liquid at 37°C temperature while 14 had fat that still contained some solids at this critical temperature. All of the Simmental carcasses had T_M well above body temperature.

The lower T_M of the fat samples from Wagyu is due to a combination of the softer fat from Wagyu (Pitchford et al., 2002) and the longer time on feed (Duckett et al., 1993). The variation of T_M within full-blood Wagyu raised within the same environment is largely due to genetic variation. There is potential for genetic improvements in fat composition by selecting Wagyu breeding stock that produce low T_M offspring.

We have developed a method for measuring T_M of edible fat, by application of proven technology on instruments that are available in all sophisticated laboratories.

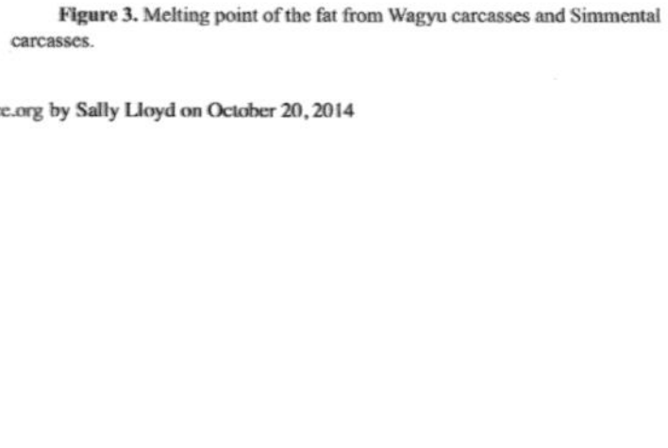

Figure 3. Melting point of the fat from Wagyu carcasses and Simmental carcasses.

Furthermore, the method described here could be encapsulated in a custom-designed apparatus with more samples and automation optimized to include the sample preparation. This method is superior to the slip T_M method and will be useful in research into improving the fat composition of meat through genetic and management strategies.

LITERATURE CITED

Adams, T. H., R. L. Walzem, D. R. Smith, S. Tseng, and S. B. Smith. 2010. Hamburger high in total, saturated and trans-fatty acids decreases HDL cholesterol and LDL particle diameter, and increases TAG, in mildly hypercholesterolaemic men. Br. J. Nutr. 103:91–98.

American Oil Chemists' Society (AOCS). 1998. Official methods and recommended practices of the AOCS. 3rd ed. AOCS Press, Champaign, IL.

Daley, C. A., A. Abbott, P. S. Doyle, G. A. Nader, and S. Larson. 2010. A review of fatty acid profiles and antioxidant content in grass-fed and grain-fed beef. Nutr. J. 9:10.

De Smet, S., K. Raes, and D. Demeyer. 2004. Meat fatty acid composition as affected by fatness and genetic factors: A review. Anim. Res. 53:81–98.

Duckett, S. K., D. G. Wagner, L. D. Yates, H. G. Dolezal, and S. G. May. 1993. Effects of time on feed on beef nutrient composition. J. Anim. Sci. 71:2079–2088.

Garmyn, A. J., G. G. Hilton, R. G. Mateescu, J. B. Morgan, J. M. Reecy, R. G. Tait, D. C. Beitz, Q. Duan, J. P. Schoonmaker, M. S. Mayes, M. E. Drewnoski, Q. Liu, and D. L. VanOverbeke. 2011. Estimation of relationships between mineral concentration and fatty acid composition of longissimus muscle and beef palatability traits. J. Anim. Sci. 89:2849–2858.

Gilmore, L. A., R. L. Walzem, S. F. Crouse, D. R. Smith, T. H. Adams, V. Vaidyanathan, X. Cao, and S. B. Smith. 2011. Consumption of high-oleic acid ground beef increases HDL-cholesterol concentration but both high- and low-oleic acid ground beef decrease HDL particle diameter in normocholesterolemic men 1, 2. J. Nutr. 141:1188–1194.

Knothe, G., and R. O. Dunn. 2009. A comprehensive evaluation of the melting points of fatty acids and esters determined by differential scanning calorimetry. J. Am. Oil Chem. Soc. 86:843–856.

Mensink, R. P., P. L. Zock, A. D. Kester, and M. B. Katan. 2003. Effects of dietary fatty acids and carbohydrates on the ratio of serum total to HDL cholesterol and on serum lipids and apolipoproteins: A meta-analysis of 60 controlled trials. Am. J. Clin. Nutr. 77:1146–1155.

Micha, R., and D. Mozaffarian. 2010. Saturated fat and cardiometabolic risk factors, coronary heart disease, stroke, and diabetes: A fresh look at the evidence. Lipids 45:893–905.

Nassu, R. T., and L. A. G. Gonçalves. 1999. Determination of melting point of vegetable oils and fats by differential scanning calorimetry (DSC) technique. Grasas Aceites 50:16–22.

Paniagua, J. A., A. G. de la Sacristana, E. Sánchez, I. Romero, A. Vidal-Puig, F. J. Berral, A. Escribano, M. J. Moyano, P. Peréz-Martinez, J. López-Miranda, and F. Pérez-Jiménez. 2007. A MUFA-rich diet improves posprandial glucose, lipid and GLP-1 responses in insulin-resistant subjects. J. Am. Coll. Nutr. 26:434–444.

Pitchford, W. S., M. P. B. Deland, B. D. Siebert, A. E. O. Malau-Aduli, and C. D. K. Bottema. 2002. Genetic variation in fatness and fatty acid composition of crossbred cattle. J. Anim. Sci. 80:2825–2832.

Uemoto, Y., T. Abe, N. Tameoka, H. Hasebe, K. Inoue, H. Nakajima, N. Shoji, M. Kobayashi, and E. Kobayashi. 2010. Whole-genome association study for fatty acid composition of oleic acid in Japanese Black cattle. Anim. Genet. 42:141–148.

Travel Report Japan November 1997

Dr. Sally Lloyd

In November 1997 I travelled to Japan as part of my PhD studies in magnetic confinement of plasma. I stayed at the National Institute of Fusion Science in Tajimi and attend the 8th International Toki Conference.

The Institute had recently finished building the largest Stellarator in operation at the time, with the ANU having a smaller version which I studied for my PhD.

Tajimi is a small city in Gifu prefecture, 30 km to the North East of Nagoya, not to be confused with Tajima prefecture where the best Kobe beef comes from. The area around Tajimi was a mixture of agricultural land and mountainous forests. Rice, vegetables and carp was grown. Many rice fields were smaller than an acre, and their position within a suburban landscape made mechanised harvest impractical. I didn't see any cows in the fields around Tajimi. Seafood was more available than beef. Milk was not popular and even the fresh milk had the boiled flavor of UHT pasturisation. Meat, seafood and vegetables were more expensive than the australian equivalents and were priced per 100 grams not per kg.

At the end of my stay I bought a fortnight's train ticket and took the opportunity to see more of the country. I travelled north, past Mount Fuji and through Tokyo to

Ibaraki prefecture, where a friend from ANU was starting a Post doc on the Japanese Tokamak, JT-60. I also travelled south to Kyushu Island to visit a friend who was teaching English to Japanese highschool students. In Kyushu were the only cattle I saw in Japan. These I now know were Akaushi grazing on the slopes of Mt Aso volcano.

Future of Biology Lies in Physics?

Some months ago Roger Stubbs raised an interesting discussion on the email having read a review in the *Daily Telegraph* (UK of *Life on the Edge: The coming of Age of Quantum Biology* by Jim Al-Khalili and Johnjoe McFadden.

We asked our in-house physics colleagues and a biologist interested in physics to provide short assessments of these developments.

Life on the Edge: The Coming of Age of Quantum Biology – A Review, By Julian Steele, PhD Scholar, School of Physics & ISEM, University of Wollongong,

To a [young] physics researcher, the discovery of quantum phenomena within biology is not surprising. Biology is, after all, a kind of applied chemistry, and chemistry is a kind of applied physics. Moreover, one could argue that the appearance of quantum phenomena in biological processes is even inevitable, given the atomic scales of most biochemical reactions. To a certain extent, this thesis is correct. Quantum mechanics is not witchcraft and has been taking place in the universe since it cooled down enough to form stable particles, and is certainly not a trick contrived by life. However, the non-trivial, indeed crucial, role quantum mechanics plays in biological processes is what's important here; from the magnetically sensitive and short-lived quantum states formed in the eyes of migratory birds, to the perplexing efficiency of photosynthesis within plants. In recognition of life as a process, and not a thing, *Life on the Edge: The coming of Age of Quantum Biology* by Jim Al-Khalili and Johnjoe McFadden prescribes an up-to-date review of the significant problems in contemporary physics, all necessary for illuminating the mystery of life. The story of quantum biology - as a scientific field - is exposed to house few real intellectual giants, experiencing strong resistance in the beginning. Of course, when heavy weights like Erwin Schrödinger were giving these ideas their first serious consideration (see, for example, Schrödinger's "What is life?"), a deep understanding of how genes, enzymes and photosynthesis worked was still missing. Early critics of quantum biology found footing in its fundamental problem; how do hot and turbulent biological systems achieve quantum coherence (the act of being in a quantum state) long enough for it to be useful? Research papers on the topic are starting to fill the pages of the world's most prestigious scientific journals, and the field is gaining real traction. Since the first evidence of quantum states existing within living matter, the initial problem has in fact now become even more puzzling, as it appears life itself is taking special measures to nurture coherence, rather than destroy it. This reveals a startling level of inherent physical "tuning", which could not be explained outside of evolutionary timescales. What life has been endowed with must be considered nothing less than incredible, as an increasing number of quantum events taking place within individual biomolecules are being shown to have far-reaching consequences for the entire organism. The origin of life itself can be explored within the framework of what is already known; the first self-replicator could have employed a search strategy capable of a quantum-level of efficiency, to conquer the game of small odds. The science of life is evolving so quickly, and on so many different fronts, it is now becoming clear that studying these phenomena will ultimately yield important discoveries. Namely, the creation of the first high-temperature quantum computer. The prospect of such rewards might be sufficient to encourage a new generation of scientists to jump ship. However, the vector of the intellectual migration seems to be - at least for the moment - pointing squarely away from physics and towards biology, as an increasing number of theoretical physicists become interested in the poetry of life.

Quantum Coherence in Biology: The key to unlocking the remaining secrets of life, or simply a cute diversion?. By Dr. Samuel Dawkins, Quantum Physicist, University of Mainz.

Legendary Physicist Erwin Schrödinger is oft-cited as one of the first to consider the cross-over of biology and quantum physics. This follows somewhat trivially from the knowledge that there is a substantial cross-over between biology and chemistry, and between chemistry and quantum physics. To understand the fundamentals of chemical interactions, one does ultimately require quantum physics. But what of the more exotic phenomena of quantum physics such as entanglement, superposition and tunnelling? Could these phenomena, which at times seem almost magical and surreal, actually lie at the heart of some of the most difficult and alluring biological puzzles such as consciousness and free will? It's an enticing thought!

But the immediate difficulty is that these quantum phenomena all broadly require there to be coherence: a special kind of order whereby particles are coordinated, thus allowing for their wave-like nature to play a role. Physicists wanting to observe and perhaps utilise a phenomenon like entanglement need to pay very special attention to coherence. 100% coherence equates to something like a perfectly unperturbed system. Any unintended noise or influence from the surrounds can lead to decoherence. With enough decoherence, behaviour converges back to what would be expected from classical principles.

For example, quantum computing attempts to harness quantum superposition, whereby particles exist in two states at the same time (up and down for example), even though they would be classically mutually exclusive. A superposition of *up and down*, however, returns to the classic *up or down*, as soon any tiny flaw with the setup creates decoherence. For this reason, most experiments on such phenomena are run close to absolute zero, thus suppressing the influence of thermal energy.

So the problem is, how could these phenomena possibly play a role in, for example, a human body, with a temperature of 310 degrees above absolute zero?! This relatively high temperature is why biological processes are sometimes said to be occurring in a molecular storm, and inherently implies an abundance of thermal noise, and therefore decoherence.

The upshot is, for something like entanglement to be possible inside the human body, there would have to be some extremely clever structure that managed to isolate the "coherent" system from the molecular storm. This does not seem to me to be a priori impossible - physicists are capable of creating such conditions, even in some cases, in structures at room temperature…. so couldn't nature achieve something similar?

This is a question I find difficult to answer - my intuition says no, but my reason says why not?

What do you think?

Last Word- The Living and the Dead By Dr. Ted Steele, CYO Foundation

The first comprehensive articulation since Schrodinger, in my view, that the "living organism is quantum coherent" was outlined in a most remarkable and original book by Dr Mai-Wan Ho first published in 1993 - *The Rainbow and the Worm- The Physics of Organisms*. Her book built on the work of German-born British solid-state physicist Herbert Frohlich. This data-based and data-rich book straddled biology and physics and went through multiple print runs, and editions, and is a science best seller – a rare achievement in itself. In my view as a connoisseur of scientific discovery, Mai-Wan Ho's book is amongst the most original I have ever read. It should have been cited in many places in *Life on The Edge* discussing the coming age of quantum biology. However it is unfortunate that it was not.

Be that as it may, Dr Mai-Wan was awarded the Ilya Prigogine Medal in 2014 for her work outlined in that book. The most striking, yet obvious fact, when you stand back, is the difference between "living" and "dead" molecular chemistry and processes – it is both common place but profoundly striking. Thus in a simple experiment Dr Ho compared the 'quantum electrodynamic' emissions of living and just killed *Drosophila* larvae. In the dead the coherent QED signals indicative of crystalline liquid chemistry (the "coloured rainbow" of the title) are completely ablated.

To this naïve observer there is something very special about the 'electrical state' of living biochemistry of supramolecular structures within living cells. e.g the quantum state of their shared electrons. So living physical chemistry is really <u>very</u>

different in its organization from the inanimate despite the physical chemistry being the same as the living.

****

Scientific Retreat Talks 2014
11th-12th April CYO'Connor Village

Quantal Structure of the Genome

EJ Steele: Introduction: The Quantal Structure of the Genome

Symposium 1 : *Lessons from the MHC*
RL Dawkins: *The Multifunctional Polymorphic Controller*
JF Williamson: *Importance conserved v deduced haplotypes*
EJ Steele : *Origins MHC polymorphism*
T Gentle: *Cytoplasmic Inheritance*
LA Smyth : *Genetics of Autoimmunity*

Symposium 2 : *Individuality*
R Davies : *Facial recognition software*
EJ Steele: *Uncertain parentage and Dam-Calf recognition*
LA Smyth: *Graft versus Host disease*

Symposium 3 : *Methods Genetic Improvement*
JF Williamson : *Genomic analysis of QTLs*
T Lloyd: *The failure of EBVs and SNPs*
SS Lloyd : *Limitations of Breedplan*
JL Valenzuela : *Experience with F1 Wagyu*
SS Lloyd: *Identification and certification by haplotype v breed*
JF Williamson: *Polling, Duplication and Paralogy*

Symposium 4 : *Wagyu and Beef Cattle*
M Buchanan : *Japanese Wagyu - example genetic diversity*
SS Lloyd: *C19 haplotype combinations and marbling*
G Wright: *The perfect steak*
M James: *Irongate Wagyu*

Marketing Beef with Low Melting Temperature Fat

Symposium 5: *Low melting point fat*
SS Lloyd : *Breeding Wagyu for low melting point (Tm) fat*
A Peggs: *Low Tm fat as marketing strategy*
M Gendy: *Potential for low Tm fat to reduce statin use*

Symposium 6: *Dairy Cattle*
JL Valenzuela : *Problems in Genetics Dairy Industry*
SS Lloyd: *Composition and melting point dairy fats*

Scientific Retreat Talks 2014
13th-14th November CYO'Connor Village

Healthy Beef & Breeding Wagyu

Roger L Dawkins: *Introduction-The Healthy Beef Project*

Symposium 1: All participants deliver 5 min introductory biographies

Symposium 2: *Certification of Breed and Pedigree*
Peter Gilmour: *ACCC inquiry into meat marketed as Wagyu*
Scott de Bruin : *Current AWA certification of Wagyu classification and pedigree*
Sally Lloyd: *C19 haplotype results for determining Breed and Pedigree*
Roger Dawkins: *Summary to date*

Symposium 3 : *What makes Wagyu meat superior?*
Veronica Nix: *Importance of marbling for taste (e.g. MSA Grading)*
Peter Gilmour : *Branded Beef Competition Tm results*
Sally Lloyd: *CYO tavern blind tasting early results*

Symposium 4: *Genetic Improvement*
Geoff Manning: *Experience with Dairy Genetic "Improvements"*
Ted Steele : *Haplotype Approach : Casein Polymorphism and Milk Production*
Mike MacLaughlin : *Dangers of Selecting for Single Dairy Traits-Getting the most out of your pasture*
Dom Bayard/Peter Gilmour : *AGBU UNE Wagyu Collaborative Genetics Research Project*
Sally Lloyd: *C19 haplotype : The role of dam in marbling*
Reports of individual results presented to those who have provided samples

Symposium 5: *Efficient Production*
Sue Lester: EBV and efficient livestock selection
Keith Hammond : Marbling on grass (Robbins Island Wagyu)
John Dawkins: Production Strategies-Grass, grain and milk

Symposium 6: *Student Papers*
Keyuri Koriya: Conserved Extended Haplotypes Human MHC
Sarah Hammond: Early–Season Leaf Removal & Phenolic Parameters of Pinot Noir

Symposium 7: *Funding Strategy*
Peter Gilmour: *Steps to AWA and MLA collaboration "Rural R&D for Profit"*
Ted Steele: *Research Support Grants&Sources*
Alan Peggs: *Potential for Western Australian solutions*
Tony Lloyd: *IP and Ownership*
Enoch Bergmann: *Realities of Commercial Testing*

Symposium 7: *Round table & interpretation of results*
Irongate Wagyu, Mayura, Robbins Island, Goorambat, MacLaughlin dairy

Visitors

During the year two visitors to the 1600 Happy Hour seminars were Professor Terry Speed , Walter and Elisa Hall Institute, Melbourne (Discussions on re-interpretation SNP data, 9th Oct 2014) and Dr Enoch Bergmann (Discussions on Healthy Beef collaboration and Pestivirus and BVDV, 10th Oct 2014 Swans Veterinary Services, Esperance).

Made in the USA
Monee, IL
05 November 2026

67378653R00040